A COLONY OF DUNG BEETLES

BY REBECCA STORM

CONTENTS

DUNG BEETLES	4
DUNG BEETLES UP CLOSE	6
ALL ABOUT DUNG	8
WHAT'S THAT SMELL?	10
MINING FOR DUNG	12
MAKING THE BALL	14
ROLL ON HOME	16
UNDERGROUND DUNG	18
MOTHER BEETLES	20
HUMANS AND DUNG BEETLES	22
TYPES OF DUNG BEETLES	24
ALL SORTS OF BEETLES	26
FUN DUNG BEETLE FACTS	28
GLOSSARY	30
INDEX	32

Copyright © 2025 Hungry Tomato Ltd

First published in 2025 by Hungry Tomato Ltd
F15, Old Bakery Studios, Blewetts Wharf, Malpas Road, Truro, Cornwall, TR1 1QH, UK.

No part of this publication may be reproduced, stored in a retrieval system, or transmitted in any form or by any means, electronic, mechanical, photocopying, recording, or otherwise, without prior written permission of the copyright owner.

A CIP catalogue record for this book is available from the British Library.

ISBN 9781835694169

Printed in China

Discover more at
www.hungrytomato.com

DISCLAIMER:
Insects are fascinating, but best to stay away! Don't touch or handle them – some insects can sting or get aggressive when they feel threatened.

Picture credits:
Abbreviations: m-middle, t-top, l-left, r-right, bg-background.

FLPA Images: 4tr, 15t, 15br, 17b, 20bl. Shutterstock: 8b; Agami Photo Agency 25mr; Almacron 7tr; Anna K Mueller 27bl; Cassandra Madsen 24bl; Daniel J.Rao 18b; D Satyajeet 25br; Graeme Shannon 23tl; Henk Bogaard 6m, Henrik Larsson 20mr; 31b; IanRedding 7br; Julija Sapic 22br, 28mr; Khlungcenter 26mr; Marieke Peche 11tr; Michael Potter11 3bl, 14br; Nick Greaves 19tr; Nik Bruining 12ml; NuayLub 21b; Olko1975 26bl; OMfotovideocontent 19b; Peter Fodor 1b, 13t; Sheris9 23br; Serge Goujon 13b; Suarez Naranjo 29tr; WH_Pics 16t, 29mr; Wondry 28bl; Valerijs Vahrusevs 27tr; Victor sopf sopf 17t; Vinicius R.Souza 24mr.

Every effort has been made to trace the copyright holders, and we apologise in advance for any unintentional omissions. We would be pleased to insert the appropriate acknowledgements in any subsequent edition of this publication.

Words in **BOLD** can be found in the glossary.

DUNG BEETLES

Dung beetles are medium-sized winged **insects**. They can be seen rolling balls of dung (animal waste) along the ground. They often fall over, giving them the nickname of "tumblebugs"!

HOW DO THEY LIVE?

A dung beetle lives in an underground burrow. Males and females usually live in separate burrows. In some **species**, the males look quite different to the females, with some having horns on their heads.

WHAT DO THEY EAT?

Dung beetles feed on the dung of large, plant-eating **mammals** that graze on grass, leaves, and shrubs.

A pair of dung beetles, male and female, from South America.

WHERE DO THEY LIVE?

Dung beetles live anywhere where cattle, buffalo, horses, antelope, and other large mammals are found.

A dung beetle in cattle dung

IT'S A BUGS WORLD

Insects belong to a group of animals known as **arthropods**. Adult arthropods have jointed legs, but do not have an inner **skeleton** made of bones. Instead, they have a tough outer "skin" called an **exoskeleton**. Most insects have at least one pair of wings.

All beetles belong to a group of bugs known as arthropods.

DUNG BEETLES UP CLOSE

The average dung beetle has six legs and a pair of flying wings. Its body is divided into three parts – head, **thorax**, and **abdomen**.

The abdomen is the largest part of the dung beetle's body. It contains the **digestive system**, and other important **organs**.

The thorax is the middle part of the body; and the legs and wings are attached here.

The head is equipped with **antennae**, eyes, brain, and mouth.

SIX LEGS PER CREEPY-CRAWLY

Beetles and other insects are sometimes called "hexapods" because they all have six legs ("hex" means "six" in Greek). This can be a bit confusing – all insects are hexapods, but not all hexapods are insects!

Dung beetles have six legs, just like all other hexapods.

Beetles have a pair of hardened wing cases that can close to cover and protect their delicate flying wings when they are on the ground.

When the beetle is flying, the wing cases unfold and stick out to help the insect fly.

The open wing cases reveal this dung beetle's flying wings.

ALL ABOUT DUNG

Dung is the solid waste material that plant-eating mammals poop out! Plant-eaters have to eat much larger amounts than meat-eaters, which means they poop much more!

For dung beetles, and many other bugs, the dung of plant-eating mammals is an important supply of food. There is always some part-digested plant material in dung, together with dead **bacteria** and useful **minerals**.

Large plant-eating animals, such as elephants, leave large piles of dung!

When dung is left on the ground, tiny plants and **fungi** from the air and soil begin to grow on it. It is these tiny plants and fungi, which will only grow in dung, that are the dung beetle's main source of food.

Mushrooms growing in cow dung

CLEAN-UP CREW

Dung beetles and other dung-eating insects are part of nature's clean-up crew. Without bugs, the world's grasslands would become covered in a layer of dung. Other insects can get rid of dead animals. Sexton beetles, for example, bury the bodies of small mammals and birds in order to feed their young!

Sexton beetle

WHAT'S THAT SMELL?

The smell of dung attracts dung beetles. The greatest number of dung beetles are found in places where there are the largest numbers of plant-eating mammals – areas of grassland such as the **Serengeti plain** in Africa.

When huge herds of mammals pass through the Serengeti plain, they leave behind lots of dung. Beetles are not short of food here!

Wildebeest grazing on the plains of the Serengeti

In scrubland, however, mammals are few and far between, and there is not much dung to be found. A fresh dropping can attract dung beetles from far away.

Beetles do not have noses. Instead, their antennae give them a superb sense of smell to find fresh poop. A dung beetle's antennae can detect the faintest smell of dung!

Scrubland in Namibia, Africa.

BEST WAY TO TRAVEL?

Like most animals, beetles normally travel headfirst. Their eyes and sense organs are located on the front of the head. This might seem very simple and obvious, but as we shall see with the dung beetle – headfirst is not always the best way to travel!

A close-up of a dung beetle's head

MINING FOR DUNG

Insects all take different approaches to a pile of dung – some tunnel straight in to find the best bits, while others are happy just to nibble at the edges. Dung beetles are not too fussy about quality!

Dung beetles are designed and equipped to dig for dung! A strong, curved shield protects the front of the head.

The edge of the head shield has notches like the teeth on a pitchfork.

The front legs have many strong spikes on them, like a comb, that are used to gather dung. Dung beetles have hooks on the ends of their legs to help them cling to the dung. They then use their bodies to press the dung into a tightly packed, solid ball.

A dung beetle in the ball-rolling process

WORKING SOLO

No matter how hard they work, dung beetles can never match the amazing achievement of termites. Termites are the mining champions of the insect world. They build mounds more than 4 m (13 feet) high, with tunnels beneath that stretch about the same distance underground. However, it takes millions of termites to build a mound, whereas each dung beetle's ball is a solo achievement.

Termite mounds in Australia

MAKING THE BALL

As a dung beetle collects more and more dung, the ball beneath its body gets bigger and bigger. Eventually the ball gets so big that the beetle is tilted forwards.

The dung beetle uses its middle legs and back legs to make the ball even bigger, by walking in a handstand position! Powerful legs help the dung beetle grip the ball and stay steady during the rolling process, even when walking backwards!

INSECT ENGINEERING

When human engineers want to measure the size, shape, and smoothness of a ball, they use an instrument called a **calliper**. The middle and back legs of the dung beetle act like two pairs of natural callipers that help the beetle to make dung balls exactly the right size.

Dung beetles use their back legs as "natural callipers"

A dung beetle using its powerful legs to roll a dung ball

Dung beetles use the hooks on their legs to get a good grip on the ball of dung. They also use them to turn the ball and make sure every bit of the surface has had the same amount of pressing between the insect's body and the ground.

Tiny hooks on their legs

ROLL ON HOME

When it's satisfied with the size and shape of its dung ball, a dung beetle can start pushing the ball straight back to its burrow. It starts walking in a straight line, ignoring the slopes and curves of the ground.

A dung beetle rolling a ball of dung in the Alor Mountains, Spain

The dung beetle makes this return journey while walking backwards, because its front legs are no good at controlling the ball. It takes all four middle and back legs to steer the ball in a straight line.

MUSCLE POWER

Although the dung beetle might seem to make very slow progress, it is usually successful – and this success shows off their impressive muscle power. When the beetle pushes a dung ball across grassland, it is equivalent to a person pushing a small car over rough ground that is covered by thick jungle!

Walking backwards in a straight line over bumpy ground is difficult, even without pushing a massive ball of dung! Dung beetles face obstacles with every journey.

A dung beetle being clumsy!

A pair of dung beetles struggling to push a ball up a hill

UNDERGROUND DUNG

Sometimes the dung ball is eaten as soon as the beetle gets back to its burrow. Often, however, the dung ball is stored in the burrow to provide food for when fresh dung is hard to find.

In many places, dung is hard to find all year round. For example, when the herds of mammals on the Serengeti **migrate** elsewhere, the amount of dung available drops dramatically.

Grassland

Fortunately for dung beetles, a rolled dung ball will stay safe to eat for several months!

Although the outside of the ball may dry out, the dung in the middle of the ball will stay damp, and the tiny fungi and plants inside will continue to grow.

Dung beetles enjoying some fresh dung

CRADLE OF DUNG

After **mating**, a female dung beetle carefully makes a special ball, choosing only the very "best quality" dung. She takes extra care on the journey home to make sure the surface of this ball is completely smooth. Inside the burrow, she lays her eggs in the middle of the ball, with just a single air hole so that the eggs can breathe. This is called a cocoon.

A dung beetle cocoon

MOTHER BEETLES

With many dung beetle species, the females stay in the burrow with their eggs, guarding them against **predators** and **parasites**. There are many bug species that feed on the eggs of other species!

If the female is successful, her eggs will soon hatch, and **larvae** will emerge. The larvae begin to feed on the tiny plants and fungi inside the dung ball.

A dung beetle larva in cow dung

They remain inside the dung until they have **pupated** and grown into their adult body shape. Then they start eating!

When the small beetles have eaten their way out of the dung ball, the patient female takes them up out of the burrow and introduces them to the wonderful world of dung-rolling!

Young beetles eat their way out of a dung ball

INSECT DEVELOPMENT

Insects develop from eggs in two different ways. With many kinds of insect, the eggs hatch into larvae that look very different from the adults. The larvae go through a stage called pupation when they change into adults. Other kinds of insect, such as cockroaches and grasshoppers, have eggs that hatch into **nymphs**, that already have the adult body shape.

Wood cockroach nymph

HUMANS AND DUNG BEETLES

The Ancient Egyptians believed the dung beetle was a very special insect. They called them scarabs, and that name is still used today. The scientific name for dung beetles is "Scarabaeidae", and they are often referred to as scarab beetles.

The Ancient Egyptians saw young beetles emerging from balls of dung and thought that this was a miraculous creation of life.

Many people wore good luck charms in the shape of scarabs, made of wood, stone, pottery, or glass. They also believed that a scarab pushing its ball across the ground represented the Egyptian gods pushing the Sun across the sky.

Dung beetles shown in Ancient Egyptian markings

In some parts of the world, dung beetles are protected by people. Signs can be found in South Africa that warn people to watch out for crossing dung beetles.

SUCCESS OR FAILURE?

In many ways, dung beetles are most useful when their efforts fail – when stored balls are not eaten, or the eggs fail to hatch. All uneaten leftover dung eventually mixes into the soil and becomes a natural fertiliser, which helps things to grow.

Dung helps grass to grow in farmer's fields.

TYPES OF DUNG BEETLES

There are thousands of species of dung beetles in different parts of the world. They all have the same basic shape and basic patterns of behaviour, but there are many differences in the details.

PRECIOUS METALS

Most dung beetles are quite dull in appearance, but some from tropical regions are very eye-catching. The precious jewel scarabs from South America can be gold or silver, and look bright and shiny!

NO ROLLING

Not every species of dung beetle collects dung balls. Aphodian dung beetles tunnel into the soil beneath large areas of dung. Once their burrows are dug, aphodian dung beetles feed whenever they want, in complete safety from predators.

LABOUR-SAVING BEETLES

Most dung beetles collect their dung from large piles left by large mammals, and have to work hard shaping each dung ball. Other dung beetles, such as the European minotaur beetle, have a more relaxed approach. They collect the dung of rabbits, which are exactly the right size and shape for dung beetles to handle.

TOO BIG TO BURY

In India, there is a species of dung beetle that rolls dung balls much larger than normal – the size of an apple or orange – and are much too big to bury! Instead, they carefully cover the ball with mud, which dries, keeping the dung fresh and damp inside.

ALL SORTS OF BEETLES

There are more species of beetles than any other insect – scientists already know of about 400,000 – and new beetle species are discovered almost every day.

FIREFLIES

Despite their name, fireflies are actually beetles! They are found in woodland and grassland throughout the world. These beetles have special organs on their abdomen that produce flashes of light. Each species has its own pattern of flashes.

WEEVILS

This is the biggest group of beetles and there are more than 40,000 known species. Weevils are also known as "snout beetles" because they have a narrow head that ends in a long snout. Weevils are plant-eating creatures.

DIVING BEETLES

Diving beetles are some of the fiercest predators. They are found in ponds and streams, and can even catch fish when they all work together!

LADYBIRD

These dome-shaped beetles are easy to identify by their distinctive spots. Some species always have the same number of spots, while others can have any number between 2 and 13. Ladybirds are popular with gardeners because they eat insect pests.

FUN DUNG BEETLE FACTS

There's so much more to know about dung beetles! Delve into some fantastic facts about these unusual creatures.

THE LIFE EXPECTANCY FOR...
most dung beetles ranges from three to five years.

DUNG BEETLES ARE QUITE A...
modern type of beetle. The oldest fossils only date back 40 million years.

SOME SPECIES SPECIALISE IN...
one particular type of animal dropping. For example, one species spends most of its time on a sloth!

DUNG BEETLES WERE...
brought into Australia by farmers to keep fly numbers low. By eating the farm animals' dung, dung beetles stopped flies overwhelming farms.

DUNG BEETLES CAN BE FOUND...

on every continent, except Antarctica!

SOME DUNG BEETLES...

eat other dung beetles' eggs as well as stealing their dung.

A DUNG BEETLE CAN...

bury 250 times its own weight in one night!

SCIENTISTS HAVE FOUND...

some dung beetles use the stars in the night sky to guide them back home when rolling dung!

SOME DUNG BEETLES...

lay their eggs inside dung that other beetles collect to eat!

GLOSSARY

Abdomen – the largest part of an insect's three-part body: the abdomen contains most of the important organs.

Antennae – a pair of special sense organs found at the front of the head on most insects.

Arthropods – bugs that have jointed legs; insects and spiders are arthropods.

Bacteria – microscopically small organisms that can live just about anywhere. Some bacteria cause disease.

Bug – one of a large number of small land animals that do not have a skeleton.

Callipers – an instrument used by engineers to measure the diameter or thickness.

Digestive system – the organs in the body that are used to process food.

Exoskeleton – a hard outer covering that protects and supports the bodies of some insects.

Fungi – (singular of *fungus*) a group of living things that are separate from plants and animals. Fungi range in size from microscopic yeasts to large toadstools and mushrooms.

Insect – a type of very small animal with six legs, a body divided into three parts, and usually two pairs of wings.

Larvae – wormlike creatures that are in the juvenile (young) stages in the life cycle of many insects.

Mammal – one of a group of warm-blooded animals that have an internal skeleton and which feed their young.

Mating – when animals join together to make babies.

Migrate – when animals move from one place to another, usually to find food or because of changes in the weather.

Minerals – natural substances found in rocks and soil that are essential for both plants and animals.

Nymphs – insects in the juvenile (young) stage in the life cycle of insects that do not produce larvae.

Organs – a part of an animal's body that performs a particular task. For example, the heart pumps blood.

Parasites – living things that live or feed on or in the body of another living thing.

Predators – animals that hunt and eat other animals.

Pupated (pupation) – the process by which insect larvae change their body shape to the adult form.

Scarabs – the Ancient Egyptian name for a dung beetles. This name is sometimes used today to refer to any dung beetle.

Serengeti plain – a large area of tropical grassland in East Africa.

Skeleton – an internal structure of bones that support the bodies of large animals such as mammals, reptiles, and fish.

Species – a group of living things that share characteristics and features.

Thorax – the middle part of an insect's body where the legs are attached.

INDEX

A
abdomen 6, 26, 30
adult insects 6
Africa 10-11, 23, 31
Ancient Egyptians 22
antennae 6, 11,
Aphodian dung beetles 24
arthropods 5, 30
Australia 13, 28

B
bacteria 8, 30
balls of dung (see *dung balls*)
beetles 4-5, 6-7, 8-9, 10-11, 12-13, 14-15, 16-17, 18-19, 20-21, 22-23, 24-25, 26-27, 28-29, 30, 31
burrows 4, 24, 16, 18-19, 20, 24
burying beetles 25

C
callipers 14, 30
clean-up insects 9
cow dung 9, 20

D
development 21
digestive system 6, 30
diving beetles 27
dung 4-5, 6-7, 8-9, 10-11, 12-13, 14-15, 16-17, 18-19, 20-21, 22-23, 24-25, 26-27, 28-29, 30, 31

dung balls 14, 15, 16, 18-19, 20, 23, 24-25

E
eggs 19, 20, 23, 29
Egyptians 22
elephant dung 8
elytra 7, 30
European minotaur beetles 25
exoskeleton 5, 30
eyes 6, 11

F
females 4, 20,
fireflies 26
fish 27, 31
flies 28
food 8-9, 18, 23, 30
fossils 28
fungi 9, 19, 20, 30

G
good luck charms 22
grassland 10, 16, 18, 26, 31

H
head 4, 6, 11, 12, 26, 30
head shield 12
hexapods 7
home 16-17, 19
hooks, legs 12, 15,
humans 22-23

L
ladybirds 27
larvae 20-21, 31
legs 6-7, 12, 14-15, 16, 30
life expectancy 29
life cycle 30

M
males 4
mammals 4-5, 8-9, 10-11, 18, 25, 30
meat-eaters 8
minerals 8, 30
mining insects 12-13
minotaur beetle 25
mud-coated dung balls 25
muscle power 16

N
nymphs 21

O
organs 6, 11, 26, 30-31

P
parasites 20, 31
plant-eaters 4, 8, 30
precious metal scarabs 24
predators 20
protecting dung beetles 23
pupation 20-21

R
rabbit dung 25
rolling dung balls 4, 13, 14-15, 16, 20

S
scarabs 22, 24
scrubland 11
sense organs 11, 30
Serengeti plain 10, 18, 31
Sexton beetles 9
shaping dung balls 14-15, 25

skeleton 5, 30-31
sloth droppings 28
smell 10-11
snout beetles 26
South America 4, 24, 28
species of dung beetle 24-25
spikes, legs 12
steering dung balls 16-17
storing dung balls 18-19

T
termite mounds 13
thorax 6, 31
'tumblebugs' 4, 17, 31

W
weevils 26
wildebeest 10
wings 5, 6-7, 30